# Earth in Space

Illustrations: Janet Moneymaker
Design/Editing: Marjie Bassler

Earth in Space
ISBN 978-1-950415-39-7

Published by Gravitas Publications Inc.
Imprint: Real Science-4-Kids
www.gravitaspublications.com
www.realscience4kids.com

I0059366

Earth is so big that it can be hard to imagine what it looks like from space.

Do you think Earth looks like one of these shapes?

Earth is a **planet.**

To be a planet, an object in space must...

1. Be big enough to have its own **gravity.**

2. Be **spherical** (ball-shaped).

3. Move around a sun.

**Gravity** is the force that holds everything to the surface of Earth.

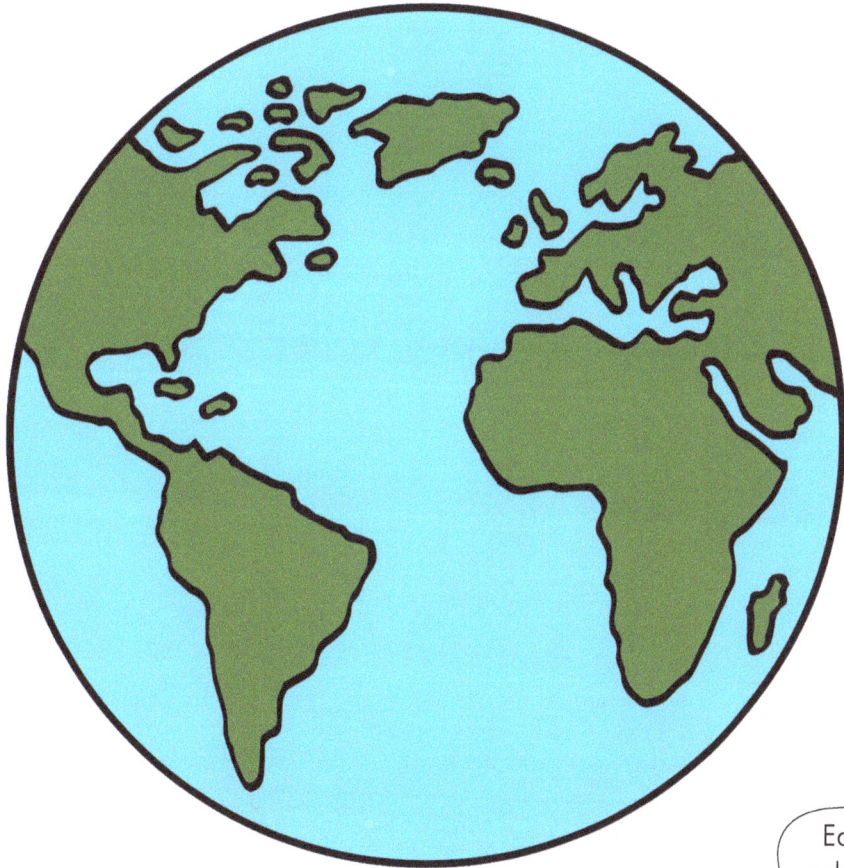

Earth is spherical.

Earth moves around the Sun in an almost circular path called an **orbit**.

Sometimes I go in circles.

Silly mouse.

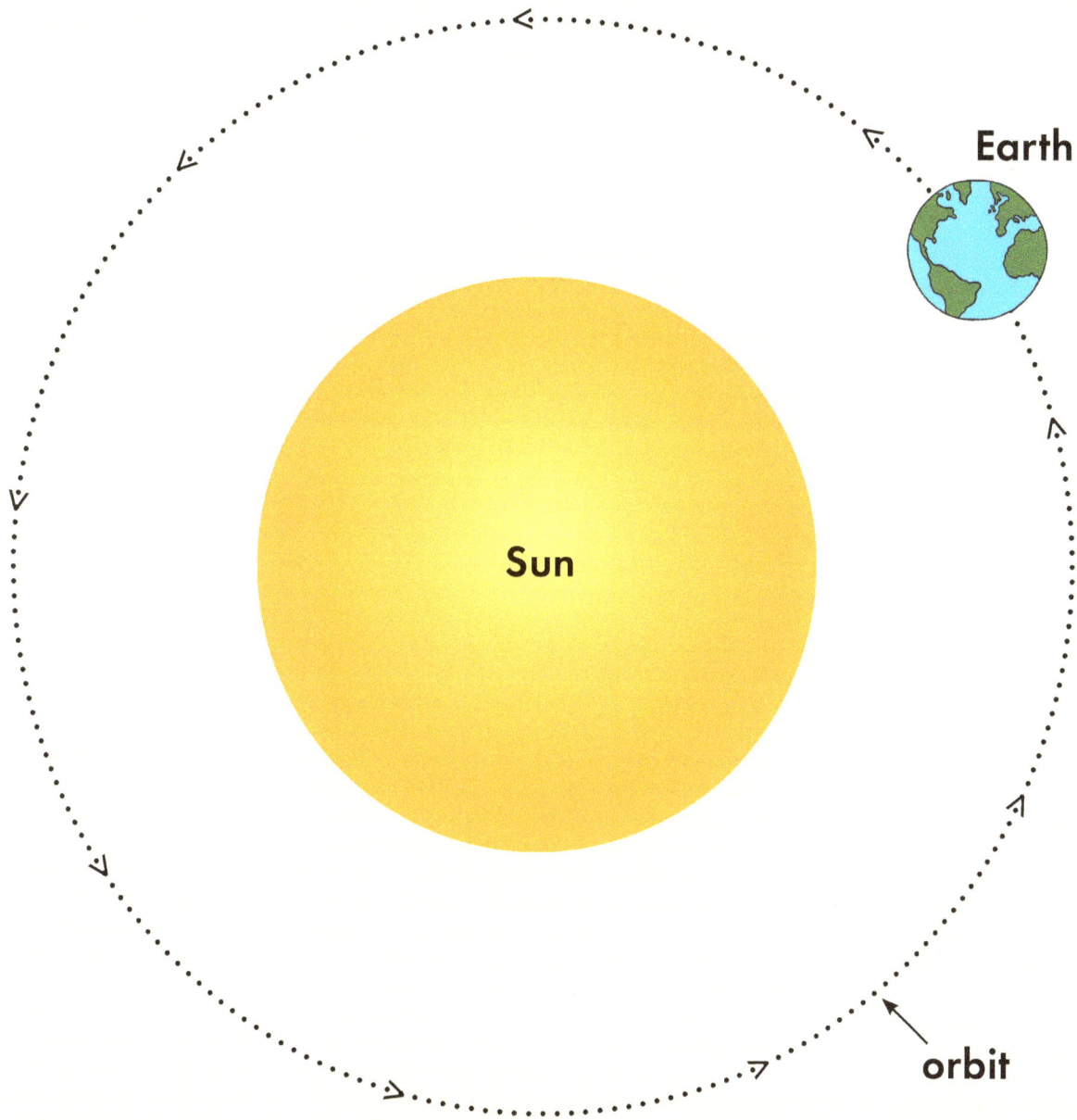

Earth also **rotates** (turns) around an **axis**. An axis is an imaginary line that goes through the center of an object.

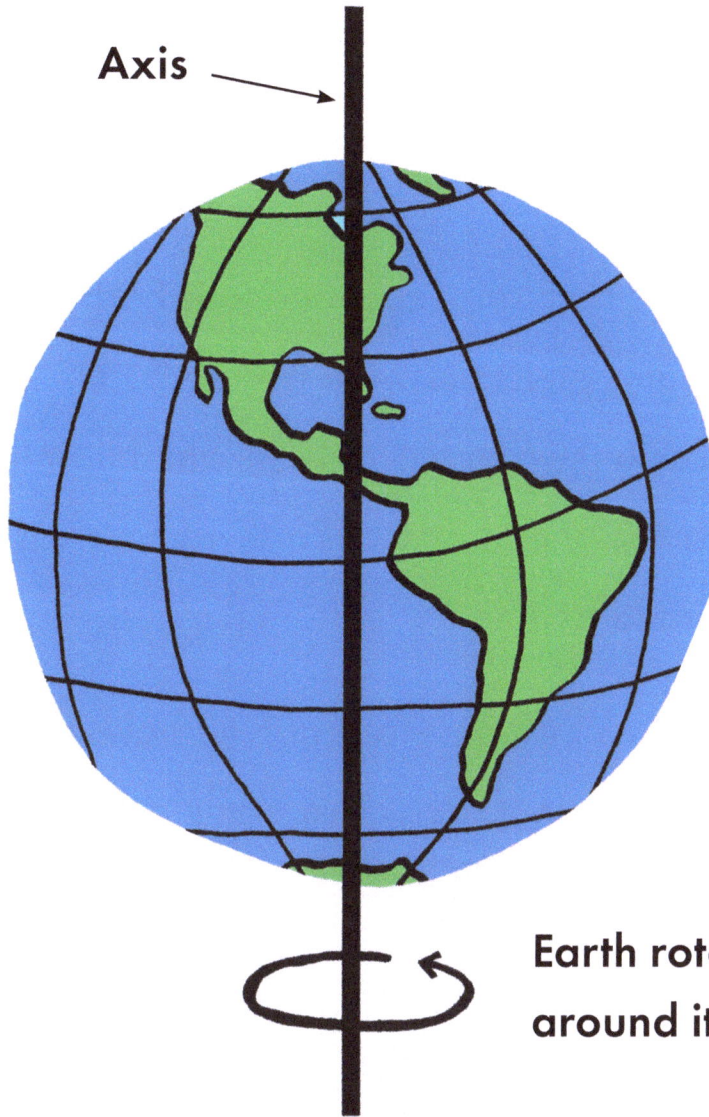

**Axis**

Earth rotates around its axis.

Day and night occur because Earth rotates on its axis. The part of Earth facing the Sun will get sunlight. Nighttime comes to that part of Earth when it has turned to face away from the Sun.

It takes 24 hours for Earth to make one full turn on its axis. This is one day.

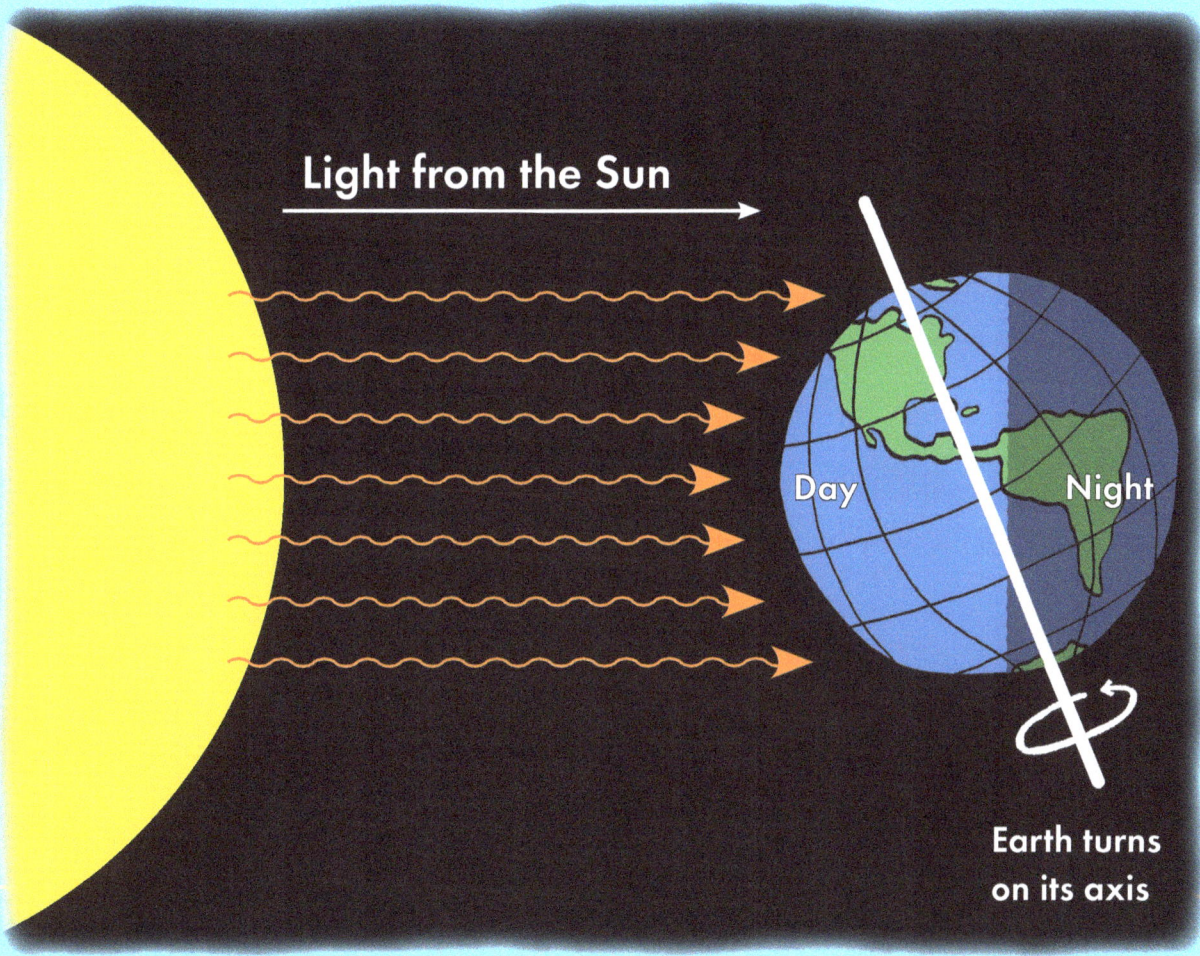

Light from the Sun

Day

Night

Earth turns
on its axis

Earth's axis is tilted instead of being straight up and down. The tilt of Earth and Earth's orbit around the Sun combine to give us the seasons—spring, summer, fall, and winter.

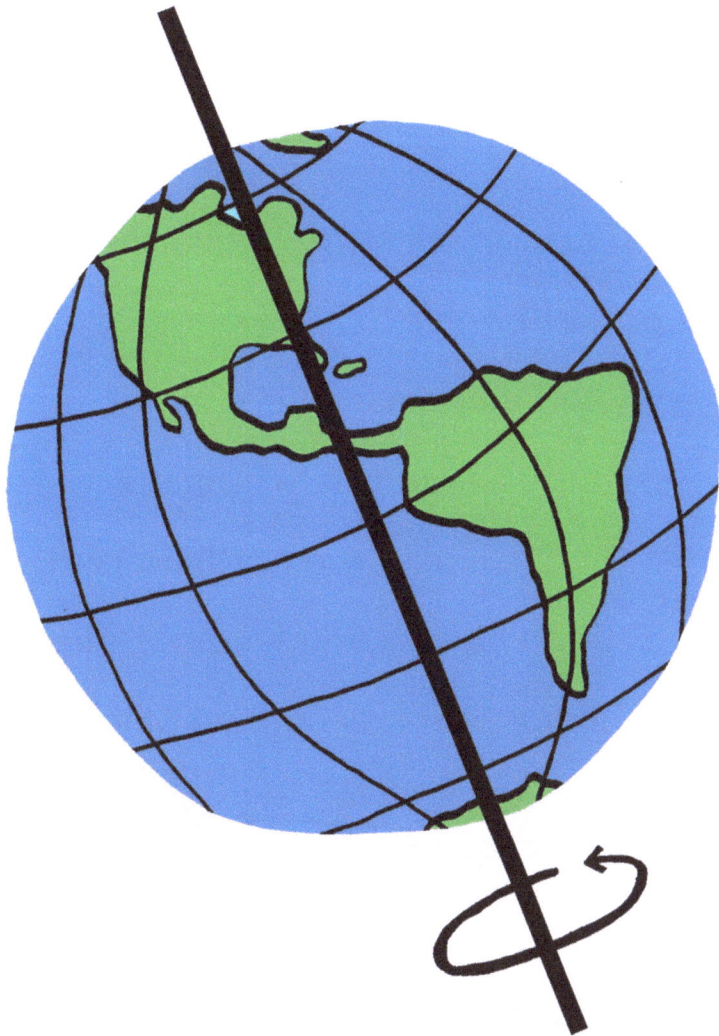

For part of the year, the northern part of Earth's axis is pointed toward the Sun and the southern part is pointed away from the Sun. This makes the northern part of Earth have warm weather. The southern part has cool weather.

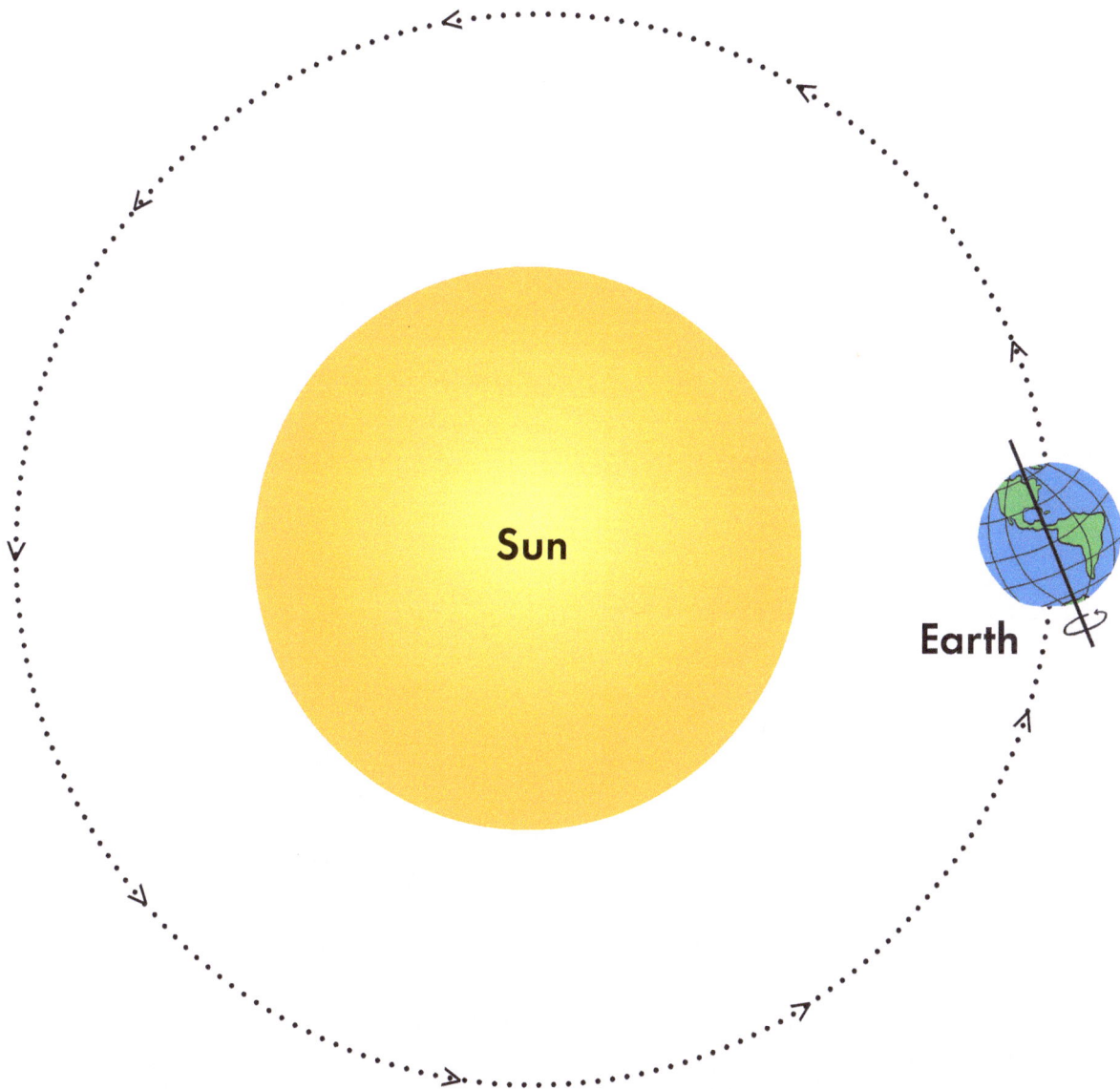

In another part of the year, Earth has moved in its orbit to a different location. Now the southern part of Earth's axis is pointed toward the Sun and the northern part is pointed away from the Sun. This makes the southern part of Earth have warm weather. The northern part has cool weather.

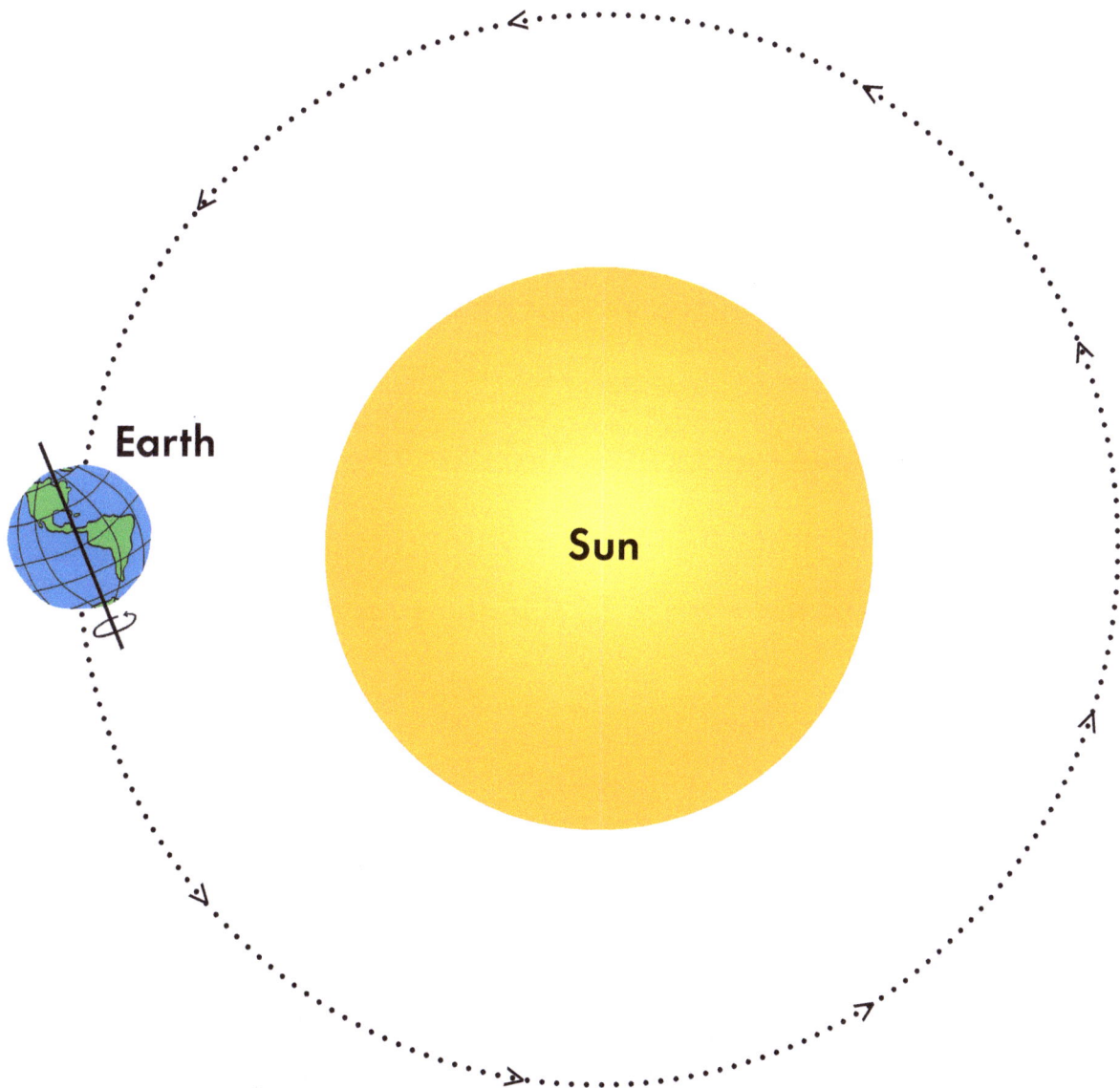

Earth

Sun

We now have machines called **satellites** that orbit Earth high above the surface. Satellites gather information and take photographs and send them back to scientists on Earth.

Earth seen from space!

Image credit: Landsat 8 satellite, Artist Concept, NASA's Goddard Space Flight Center

Earth's place in space and its relationship to other objects in space are big topics. Much has been learned, and there is much more left to discover.

I want to go to Mars!

I like it here.

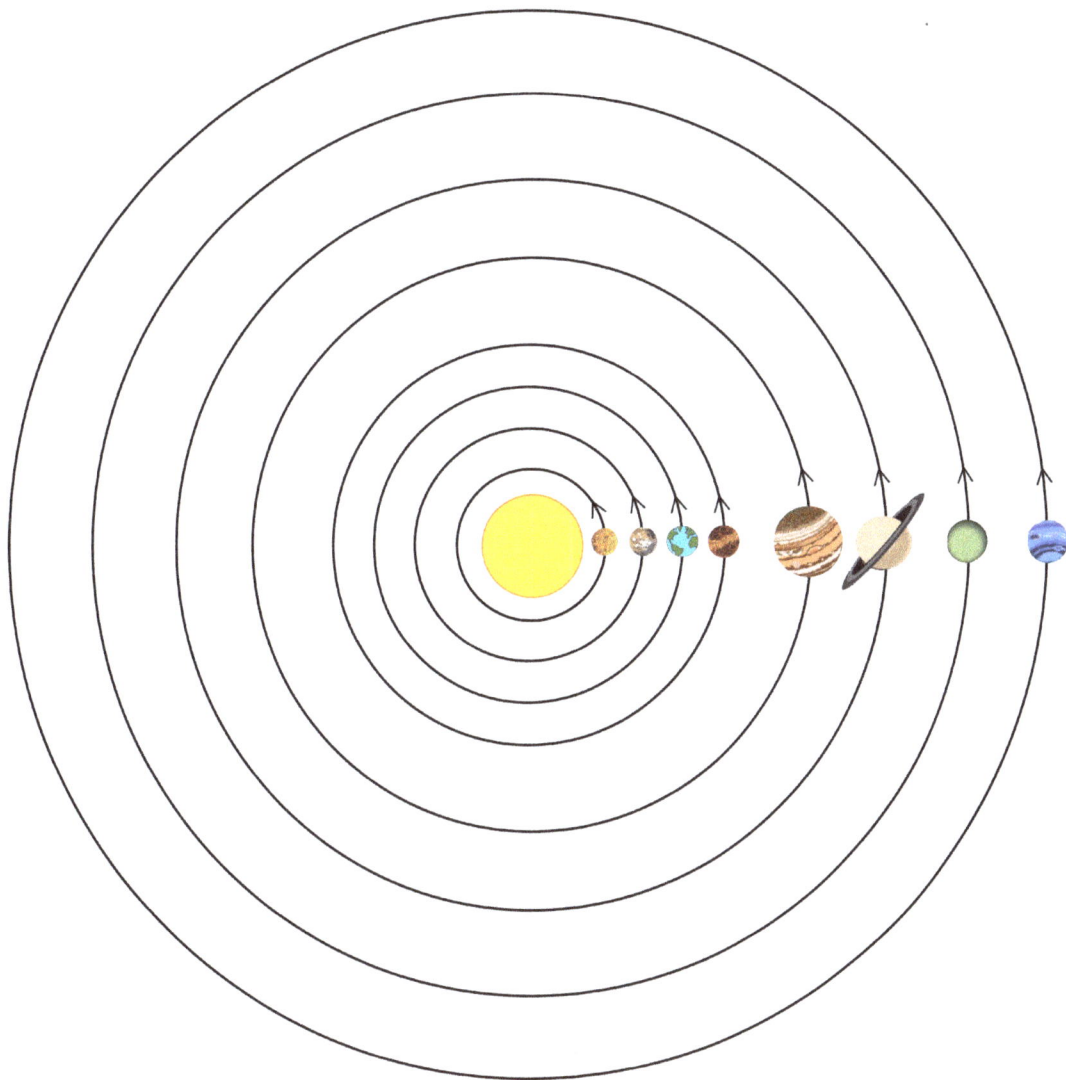

Earth is one of 8 planets that orbit the Sun.

# How to say science words

**astronomer**  (uh-STRAH-nuh-mer)

**astronomy**  (uh-STRAH-nuh-mee)

**axis**  (AK-suhs)

**Earth**  (ERTH)

**gravity**  (GRA-vuh-tee)

**imaginary**  (i-MAA-juh-ner-ee)

**orbit**  (AWR-buht)

**planet**  (PLA-nuht)

**rotate**  (ROH-tayt)

**satellite**  (SA-tuh-liyt)

**science**  (SIY-uhns)

**scientist**  (SIY-uhn-tist)

**space**  (SPAYSS)

**spherical**  (SFIR-i-kuhl)

# What questions do you have about EARTH IN SPACE?

# Learn More Real Science!

## Complete science curricula from Real Science-4-Kids

## Focus On Series

**Unit study** for elementary and middle school levels

**Chemistry**
**Biology**
**Physics**
**Geology**
**Astronomy**

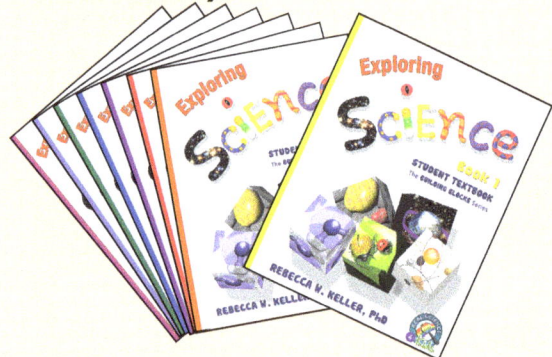

## Exploring Science Series

**Graded series** for levels K–8. Each book contains 4 chapters of:

**Chemistry**
**Biology**
**Physics**
**Geology**
**Astronomy**

www.ingramcontent.com/pod-product-compliance
Lightning Source LLC
Chambersburg PA
CBHW040149200326
41520CB00028B/7543